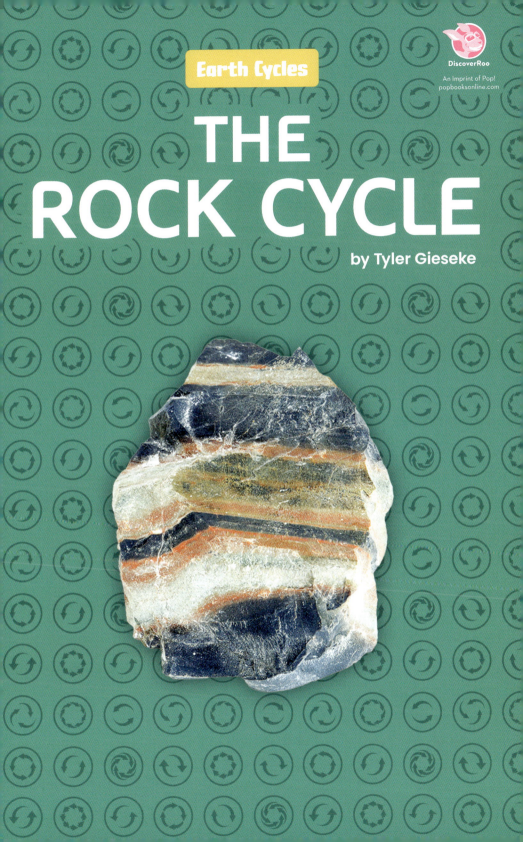

abdobooks.com

Published by Pop!, a division of ABDO, PO Box 398166, Minneapolis, Minnesota 55439. Copyright © 2023 by Abdo Consulting Group, Inc. International copyrights reserved in all countries. No part of this book may be reproduced in any form without written permission from the publisher. DiscoverRoo™ is a trademark and logo of Pop!.

Printed in the United States of America, North Mankato, Minnesota.

052022
092022

THIS BOOK CONTAINS RECYCLED MATERIALS

Cover Photo: Shutterstock Images
Interior Photos: Shutterstock Images
Editor: Elizabeth Andrews
Series Designer: Laura Graphenteen

Library of Congress Control Number: 2021951849

Publisher's Cataloging-in-Publication Data

Names: Gieseke, Tyler, author.
Title: The rock cycle / by Tyler Gieseke
Description: Minneapolis, Minnesota : Pop, 2023 | Series: Earth cycles | Includes online resources and index
Identifiers: ISBN 9781098242220 (lib. bdg.) | ISBN 9781098242923 (ebook)
Subjects: LCSH: Geochemical cycles--Juvenile literature. | Rock cycles--Juvenile literature. | Geological cycles--Juvenile literature. | Seismological cycles--Juvenile literature. | Earth sciences--Juvenile literature. | Environmental sciences--Juvenile literature.
Classification: DDC 550.0--dc23

Pop open this book and you'll find QR codes loaded with information, so you can learn even more!

Scan this code* and others like it while you read, or visit the website below to make this book pop!

popbooksonline.com/rock

*Scanning QR codes requires a web-enabled smart device with a QR code reader app and a camera.

TABLE OF CONTENTS

CHAPTER 1
Three Kinds of Rock............... 4

CHAPTER 2
Igneous Rock..................... 12

CHAPTER 3
Sedimentary Rock................. 18

CHAPTER 4
Metamorphic Rock 24

Making Connections.............. 30
Glossary 31
Index........................... 32
Online Resources................ 32

CHAPTER 1

THREE KINDS OF ROCK

A volcanic **eruption** sends hot gases and ash soaring into the sky. **Lava** also emerges from the volcano, sometimes exploding up and sometimes oozing out. As the lava flows down the volcano, it cools and hardens. The lava is now a rock.

WATCH A VIDEO HERE!

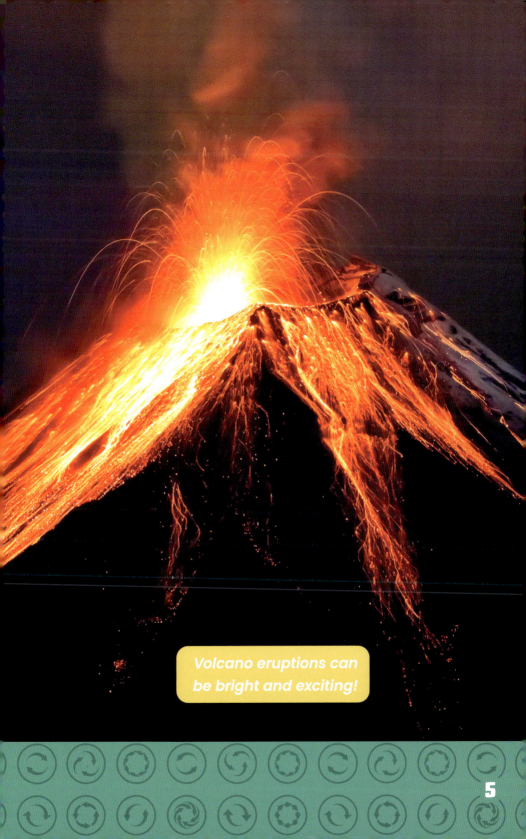

Volcano eruptions can be bright and exciting!

This rock's journey is just beginning. Over many years, wind and rain break off tiny pieces from it. These pieces collect in a low place. Over a long time, more small

Running water can wear down rock and soil.

pieces of rock pile on top. Once they're buried deep enough, the pieces on the bottom squish together. They form a new kind of rock.

Sand and other small pieces of rock continue to pile on top of the new rock. The new rock is slowly buried deeper and deeper in the earth. Perhaps millions of years later, it ends up miles beneath Earth's surface. There, incredible heat and **pressure** change it into a new rock, again!

DID YOU KNOW? Tamu Massif is the world's largest volcano. It is underwater in the Pacific Ocean!

This small island near Portugal is the top of an old underwater volcano.

ONE ROCK'S JOURNEY

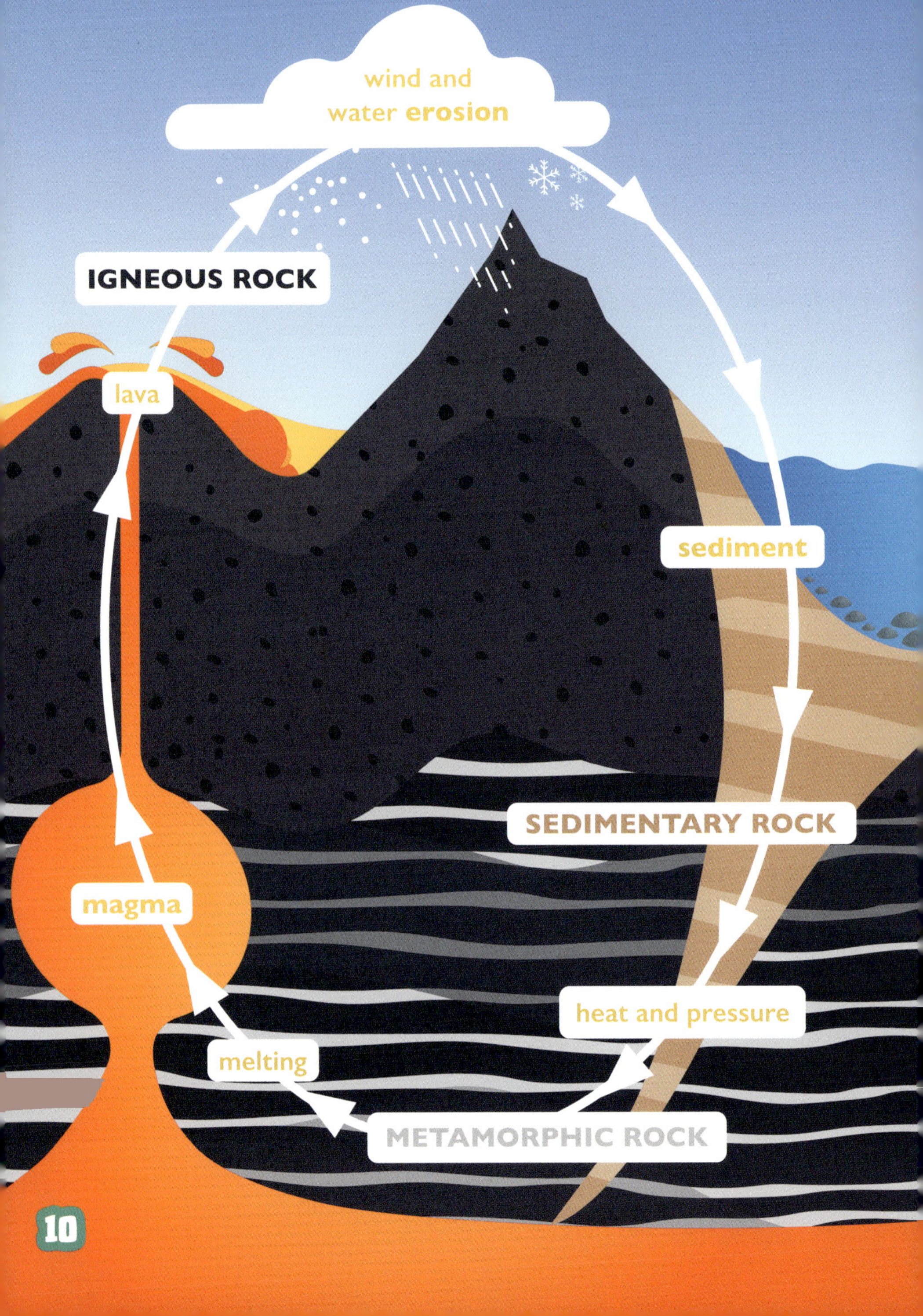

wind and water **erosion**

IGNEOUS ROCK

lava

sediment

SEDIMENTARY ROCK

magma

heat and pressure

melting

METAMORPHIC ROCK

The journey of this rock and all other rocks from one form to another is called the rock cycle. The rock cycle is an important Earth cycle that forms and shapes the land. There are three kinds of rock: igneous rock, sedimentary rock, and metamorphic rock. Each type of rock can turn into either of the other two! Together, the three types make up the rock cycle.

IGNEOUS **SEDIMENTARY** **METAMORPHIC**

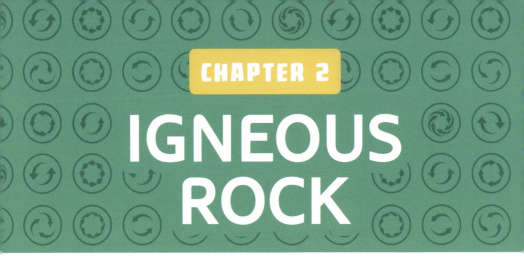

CHAPTER 2
IGNEOUS ROCK

When hot **magma** cools and becomes solid, it is called igneous rock. This kind of rock can form above ground or deep under the ground.

LEARN MORE HERE!

igneous rock

A lava field cools to form igneous rock.

Basalt in Northern Ireland

Igneous rocks are often created by volcanoes. **Lava** from volcanoes cools to form igneous rocks such as basalt and obsidian. Basalt is a dark rock used in bricks, walls, and even railroad tracks. Obsidian is another dark rock that looks like glass. Ancient peoples used obsidian in tools and weapons.

obsidian

Igneous rock can also form when magma deep in the Earth slowly cools. These rocks can have visible sections of **minerals**. Granite is a good example of this type of igneous rock.

DID YOU KNOW? Most of the rock that makes up Earth is igneous rock.

Tourists can see lots of igneous rock at Hawai'i Volcanoes National Park. The black rock there is made of lava that quickly cooled. In fact, the Hawaiian Islands themselves were formed from volcanic activity.

Kilauea Volcano spills lava into the ocean at Hawai'i Volcanoes National Park.

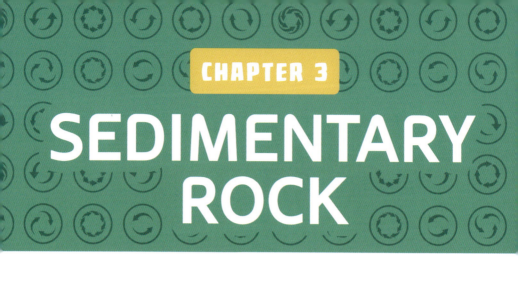

CHAPTER 3
SEDIMENTARY ROCK

Sedimentary rock is made of sand and other small pieces of rock that became buried and squished together. **Erosion** from wind and water breaks igneous or metamorphic rocks into small pieces.

EXPLORE LINKS HERE!

When they are buried deep under more **sediment**, **pressure** builds up and pushes the pieces together.

Sedimentary rock can look like art.

coal

Sandstone is one type of sedimentary rock. It is used in buildings, in pavement, and to make glass. Coal is a sedimentary rock made from pieces of dead plants and animals. Coal is used as a fuel.

Limestone is a third type of sedimentary rock. People have used it from ancient times. For example, the Egyptian pyramids at Giza are made of limestone blocks. Limestone can also be used to make cement.

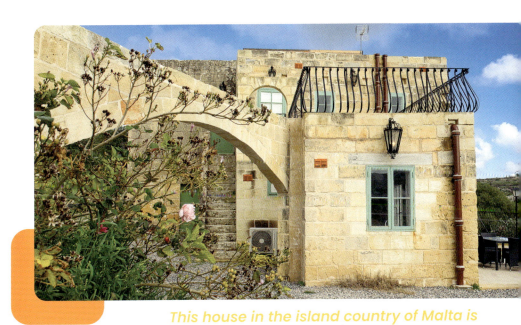

This house in the island country of Malta is made of sandstone.

GRAND CANYON FORMATION

Millions of people visit the Grand Canyon each year.

The story of the Grand Canyon begins hundreds of millions of years ago. Layers of sediment stacked up over a long time. Sometimes the area was covered with water. At other times, it was dry like a desert. In every period, bits of rocks formed layers of sediment.

The Colorado River began to flow over the sedimentary rock about 5 or 6 million years ago. Over time, the river carved a path through it. This created the canyon's depth.

The Grand Canyon in Arizona is made of sedimentary rock. The different colors of stone are from sediment getting squished together to form rock at different times. Scientists can study these layers to learn what the area was like when each section formed.

DID YOU KNOW? Scientists think the youngest layer of rocks in the Grand Canyon is 270 million years old!

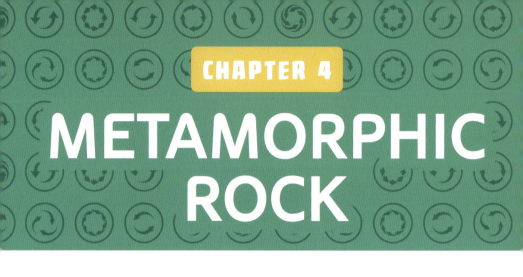

CHAPTER 4
METAMORPHIC ROCK

Metamorphic rock is the third type of rock in the cycle. When igneous or sedimentary rock becomes buried deep in the earth, it faces intense heat and **pressure**. These **forces** change the rock into metamorphic rock.

COMPLETE AN ACTIVITY HERE!

When granite is pressed very hard, it transforms into gneiss, a metamorphic rock. Gneiss has lines in it. The lines show how the granite folded together during its transformation. Another metamorphic rock is marble, which people use in countertops and fancy buildings.

Builders use gneiss for floors, gravestones, and more.

Metamorphic rock that comes to the Earth's surface can face **erosion** and later form a sedimentary rock. Or, metamorphic rock can melt beneath the Earth's surface into **magma** and later cool to form an igneous rock.

The rock cycle goes on and on. The igneous rocks, sedimentary rocks, and metamorphic rocks around us make

DID YOU KNOW? Metamorphic rock in California's Marble Canyon looks like zebra stripes.

Many statues are made of marble.

Earth a special place to live. The rock cycle is important, but it is just one of the many Earth cycles that shape our world.

ROCK CYCLE TRANSFORMATIONS

Some Earth cycles have steps that go in order. The steps in those cycles go in a circle and repeat. But the rock cycle is different. Each type of rock in the rock cycle can change into either of the other two types of rock. So, the steps of the rock cycle make a web, not a circle! Put your finger on one of the rocks on this page. Trace the arrows and try to remember how the rock changes each time. If you need help, look at page 29.

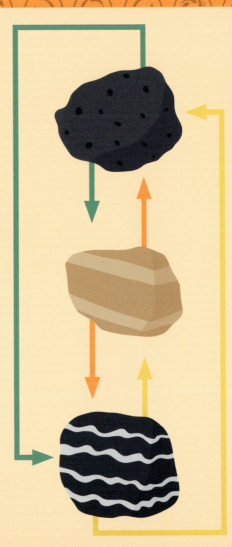

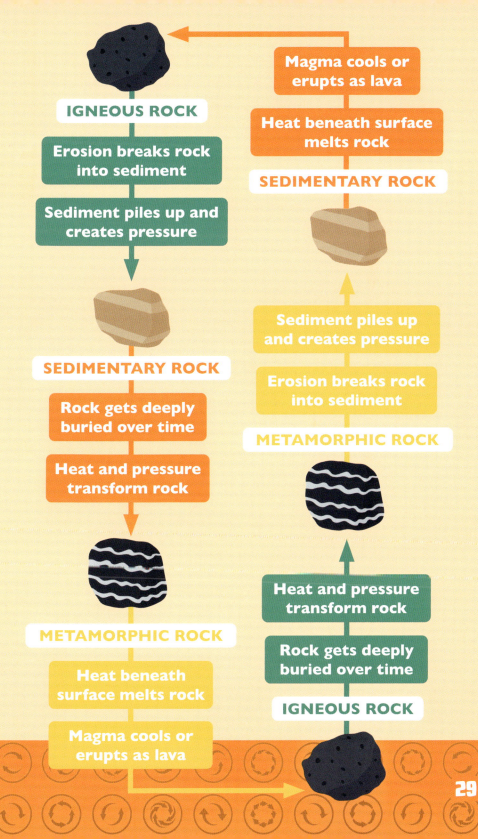

MAKING CONNECTIONS

TEXT-TO-SELF

Which of the three types of rock is your favorite? Why?

TEXT-TO-TEXT

What other books have you read about rocks? Were there facts in this book that were also in those books?

TEXT-TO-WORLD

Would you want to visit the Grand Canyon or the Hawaiian volcanoes someday? Why or why not?

GLOSSARY

erosion — the slow wearing away of something, often rock or soil.

eruption — when something bursts suddenly.

force — a push or pull.

lava — melted rock above Earth's crust.

magma — melted rock beneath Earth's crust.

minerals — any materials, such as copper or iron, that are not from living things.

pressure — a steady push against something.

sediment — material such as dirt that is moved and put down by water or wind.

INDEX

basalt, 15

gneiss, 25

Grand Canyon, 22–23

granite, 16, 25

Hawai'i, 17

lava, 4, 10, 15, 17, 29

limestone, 21

marble, 25

obsidian, 15

sandstone, 20

ONLINE RESOURCES
popbooksonline.com

Scan this code* and others like it while you read, or visit the website below to make this book pop!

popbooksonline.com/rock

*Scanning QR codes requires a web-enabled smart device with a QR code reader app and a camera.